WEATHER

Anita Chapman

SERIES EDITORS
Jim Cummins • David Freeman • Yvonne Freeman
Les Asselstine • Catherine Little

ADVISORY BOARD

Michelle Beauregard
Consultant, Diverse Learning, ELL
Calgary Catholic School District
Alberta

Carol Berndt
English Language Learners,
Learning Support Services
Calgary Board of Education
Alberta

Wilfred Burton
SUNTEP (Saskatchewan Urban Native
Teacher Education Program) Faculty
University of Saskatchewan
Saskatchewan

Beth Gunding
ESL/ELD Consultant
Ontario

Janice Lam
ELL Consultant
Vancouver Board of Education
British Columbia

Sandra Mills-Fisher
Instructional Leader
Toronto District School Board
Ontario

Suzanne Muir
Equity and Inclusive Education
Coordinator
Halton District School Board
Ontario

Sharon Newmaster
ESL/ELD Consultant
Waterloo Region District School Board
Ontario

Sharon Seward
Consultant, Diverse Learning, ELL
Calgary Catholic School District
Alberta

Diane Tijman
District Curriculum Coordinator,
ESL and Multiculturalism
Richmond School Board No. 38
British Columbia

Heather Vallee
Curriculum Consultant, ESL/ELD
York Region District School Board
Ontario

PEARSON

Pearson Canada Inc.
26 Prince Andrew Place
Don Mills, ON M3C 2T8
Customer Service: 1-800-361-6128

Rubicon
www.rubiconpublishing.com

© 2014 Rubicon Publishing Inc. Published by Rubicon Publishing Inc. All rights reserved. No part of this publication may be reproduced or transmitted in any form or by any means, electronic or mechanical, including photocopying, recording, taping, or any information storage and retrieval system, without the prior written permission of the copyright holder.

Big Idea is a trademark of Rubicon Publishing Inc.

Every reasonable effort has been made to trace the owners of copyrighted material and to make due acknowledgement. Any errors or omissions drawn to our attention will be gladly rectified in future editions.

Associate Publisher: Cheryl Turner
Editors: Mariana Aldave, Stephanie Leggett
Editorial Assistant: Vicky He
Creative Director: Jennifer Drew
Designer: Brandon Koepke

13 14 15 16 17   5 4 3 2 1

ISBN: 978-1-77058-738-0

Paper used in the production of this book is a natural, recyclable product made from wood grown in sustainable forests. The manufacturing process conforms to the environmental regulations of the country of origin.

We acknowledge the financial support of the Government of Canada through the Canada Book Fund for our publishing activities.

CONTENTS

4   INTRODUCTION:
Exploring Weather

6   CHAPTER 1:
Weather in the Air

- **8** Air and Weather
- **10** Water and Weather
- **12** *Make It Rain*
- **13** *Interpret a Weather Forecast*

14   CHAPTER 2:
Weather Patterns

- **16** Seasons and Climate
- **18** Measuring Weather
- **20** *Make a Windsock*
- **21** *Keep a Weather Log*

22   CHAPTER 3:
Weather and Our World

- **24** Weather and Living and Non-living Things
- **26** Extreme Weather
- **28** *Climate Change*
- **29** *Test Cloth*

- **30** CONNECTIONS
- **32** GLOSSARY
- **32** INDEX

BIG IDEA

Exploring Weather

BIG IDEA — Weather is important to people and their living and non-living environment.

- lightning
- hail
- tornado
- sun
- wind
- cloud

Weather happens in the air around us. Some days the weather is sunny and warm. Some days the weather is windy and cold. There are many different types of weather.

snow

sleet

rain

How does weather change your world?

CHAPTER 1

WEATHER IN THE

Earth has a blanket of air around it. The air around Earth is called the atmosphere. Weather happens in the atmosphere.

The sun's energy heats the atmosphere. You can feel the heat energy from the sun in the air around you. Hot and cold air are part of weather.

ATMOSPHERE

Earth

FYI

Air that is close to Earth is warmer than air high above Earth.

6

AIR

sun

BIG IDEA — Weather happens in the air around us.

You can use your senses to **observe** the weather.

What can you see?

What can you hear?

What can you feel?

1. List three words that tell what the weather is like outside today.

2. Which sense do you use most often to observe the weather?

7

Air and Weather

BIG IDEA Moving air causes changes in the weather.

AIR PRESSURE

Air pressure is the weight of air pushing down on Earth.
You do not feel air pressure because it is around you all the time.

Air is made up of small air particles.

High Air Pressure

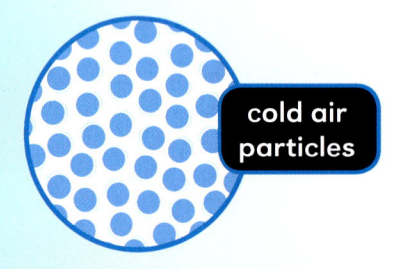

cold air particles

cold air

warm air

In cold air, the air particles move slowly and are close together.
This makes cold air heavy.
Air goes down when it cools.

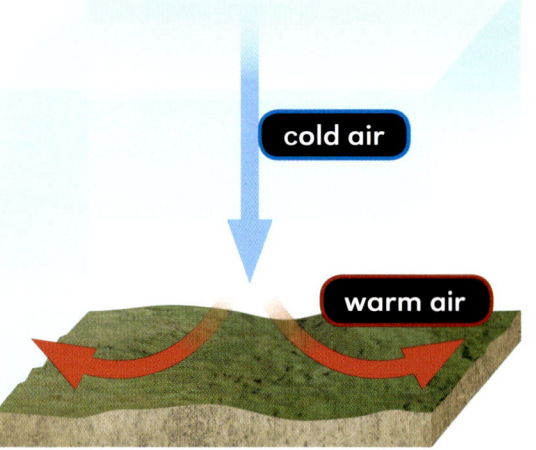

Low Air Pressure

warm air particles

warm air

cold air

In warm air, the air particles move quickly and are far apart.
This makes warm air light.
Air goes up when it warms.

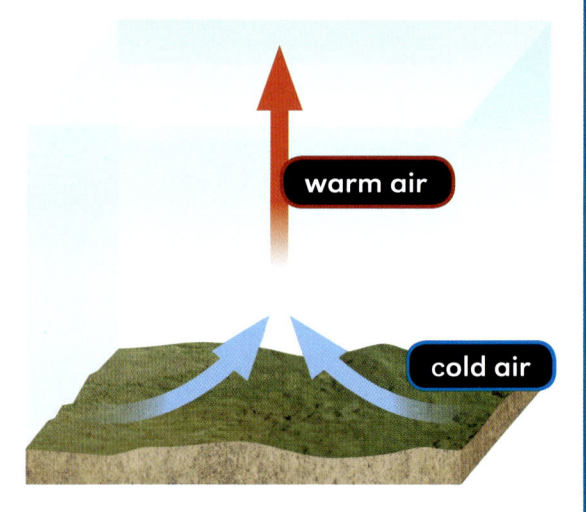

WIND

Air moves from an area with high air pressure to an area with low air pressure. This makes wind. When there is a big difference in air pressure, the wind is very strong.

high air pressure | sun | low air pressure
wind
water | land

Water and land are heated by the sun during the day.

Water heats up more slowly than land. The cool air over the water (high air pressure) moves onto the land (low air pressure). The wind moves from the water onto the land.

Air Pressure Systems

A large area of air with the same air pressure is called an air pressure system. Air pressure systems move around and make the weather change.

A high air pressure system usually brings sunny weather.

A low air pressure system usually brings clouds and rain or snow.

high air pressure

low air pressure

1. What is the weather like when the air pressure is high?
2. What is the weather like when the air pressure is low?
3. What causes wind?
4. Think about a house with two levels. Which level would be hotter in the summer?

FYI

A wind gets its name from where it comes from. A north wind comes from the north.

Water and Weather

BIG IDEA: Water moves through the water cycle. This causes changes in the weather.

THE WATER CYCLE

Most of Earth is covered by water.
Water moves from Earth into the air.
When the air is warm, the water falls back to Earth as rain.
When the air is cold, the water falls back to Earth as snow, hail, or sleet.

Rain, snow, hail, and sleet are called precipitation.

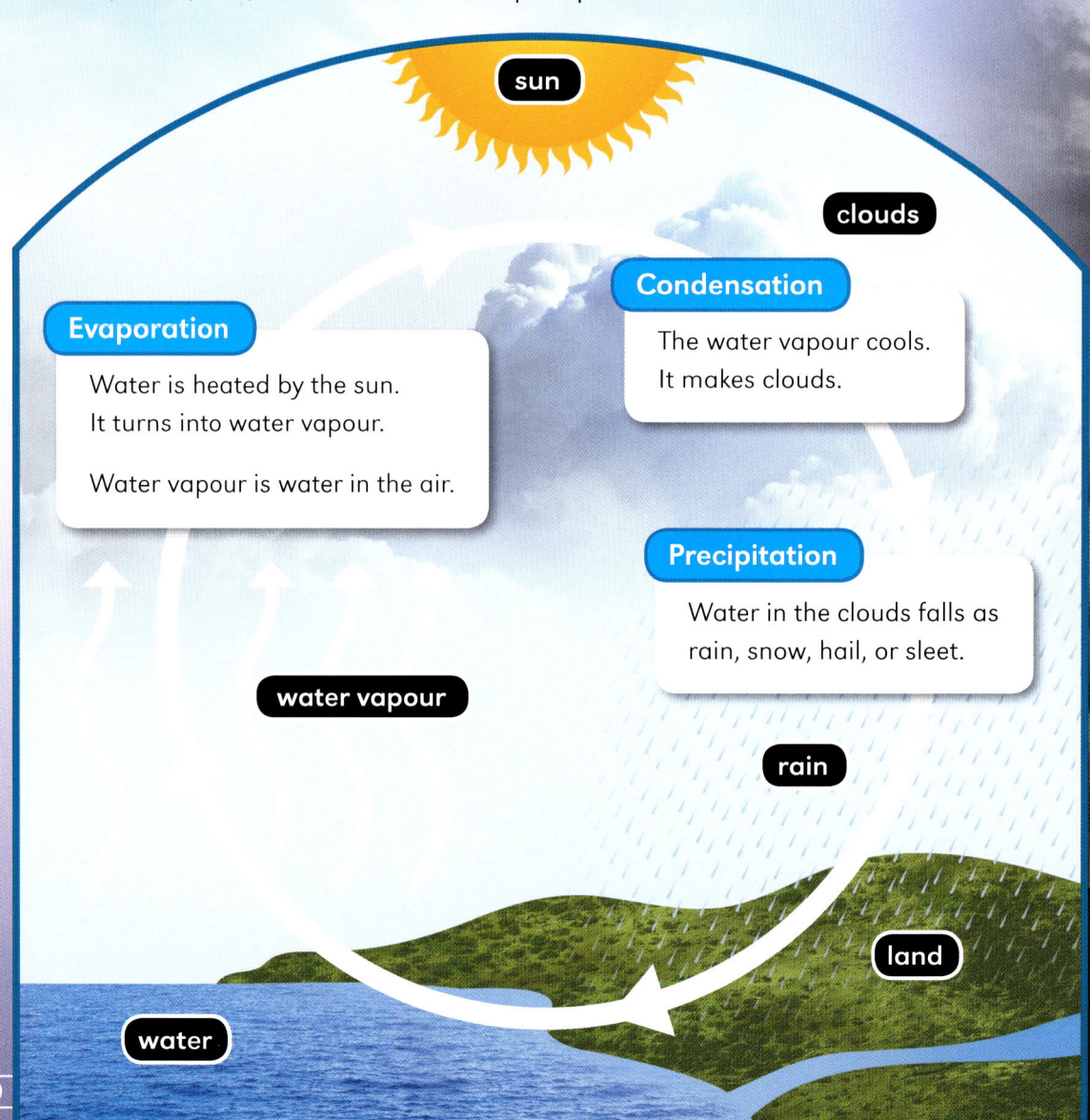

sun

clouds

Evaporation
Water is heated by the sun.
It turns into water vapour.

Water vapour is water in the air.

Condensation
The water vapour cools.
It makes clouds.

Precipitation
Water in the clouds falls as rain, snow, hail, or sleet.

water vapour

rain

land

water

CLOUDS

There are different types of clouds. Sometimes clouds are high in the sky. Sometimes clouds are low in the sky. The amount of sky covered by clouds is called cloud cover.

Clouds can help people **predict** the type of weather that is coming.

Cirrus Clouds

Cirrus clouds are high in the atmosphere where it is cold. If you see cirrus clouds, the weather might be changing.

Cumulus Clouds

Cumulus clouds are in the middle of the atmosphere. If you see cumulus clouds, it is likely going to be a nice day.

Cumulonimbus Clouds

Cumulus clouds can change into cumulonimbus clouds. If you see cumulonimbus clouds, a thunderstorm might be coming.

Stratus Clouds

Stratus clouds are low in the atmosphere. If you see stratus clouds, precipitation might be coming.

FYI

The stories of the Haida and some other First Nations tell of the Thunderbird that brings thunder and lightning.

1. What happens in the water cycle?
2. What type of clouds can tell you that rain might be coming?
3. Look at the clouds outside. What type of weather might be coming later today? Why do you think that?

CHAPTER 1 — WEATHER IN THE AIR

ACTIVITY 1: MAKE IT RAIN
Make a model of the water cycle.

YOU WILL NEED: glass jar, hot water, food colouring, ice cubes, glass plate, pencil and paper

1. Pour 5 cm of hot water into the glass jar. Add food colouring.

BE SAFE
The water has to be hot. Ask your teacher for help.

2. Cover the jar with the plate. Wait a few minutes.

3. Look in the jar. What changes do you see?

4. What you see is called an observation. Record your observations.

5. Put the ice cubes on the plate. Wait a few minutes.

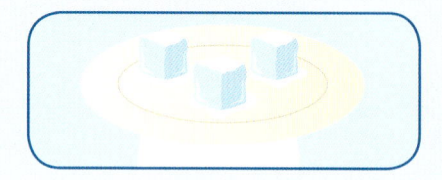

6. Record your observations.

7. Lift the plate about 10 cm.

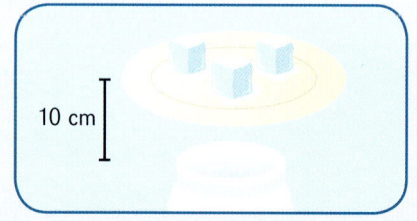

8. Record your observations.

ANALYZE & REFLECT

1. What happened to the water in the jar? Why?
2. How did the ice cubes help to make it rain?
3. Draw a picture of your model of the water cycle. Label evaporation, water vapour, condensation, and precipitation.

ACTIVITY 2: INTERPRET A WEATHER FORECAST

Meteorologists are scientists who study weather. They use weather measurements to predict the weather. Predicting the weather is called making a weather forecast.

Day	High Temperature	Low Temperature	Cloud Cover and Precipitation
Monday, 4 February	0°C	-5°C	Cloudy periods
Tuesday, 5 February	2°C	-3°C	Cloudy with rain
Wednesday, 6 February	1°C	-4°C	Mainly cloudy
Thursday, 7 February	-5°C	-9°C	Sunny
Friday, 8 February	-1°C	-7°C	Cloudy with snow
Saturday, 9 February	-3°C	-7°C	Cloudy periods
Sunday, 10 February	-2°C	-6°C	Sunny

1. Read the weather forecast in the chart above.

2. Which day is going to be the coldest?

3. Which day is going to be the warmest?

4. What is the weather forecast for Saturday? What type of clothing would you wear?

5. Look at the weather forecast for Friday. What are some activities that you could do after school in this weather?

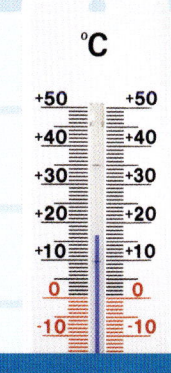

ANALYZE & REFLECT

1. Why are the air temperatures so cold in this forecast?
2. Which days could have snow? Why?
3. Which day has the type of weather that you like best? Why?
4. How would this forecast be different in July?

13

CHAPTER 2

WEATHER PAT

Different places have different weather. Different seasons bring different weather. Different clouds can bring different types of weather.

These are weather patterns. Meteorologists use patterns to help them predict the weather.

Summer
Yukon

Spring
British Columbia

Fall
Alberta

Spring
Manitoba

14

TERNS

BIG IDEA Weather patterns help us predict the weather.

- Winter — Saskatchewan
- Fall — Prince Edward Island
- Winter — Ontario
- Summer — Newfoundland

?
1. What season is it right now where you live? List five things that show what season it is.
2. Why do meteorologists look for weather patterns?

15

Seasons and Climate

BIG IDEA — Weather and climate are different around the world.

SEASONS

Earth travels around the sun once every year. As Earth travels, places around the world have different seasons.

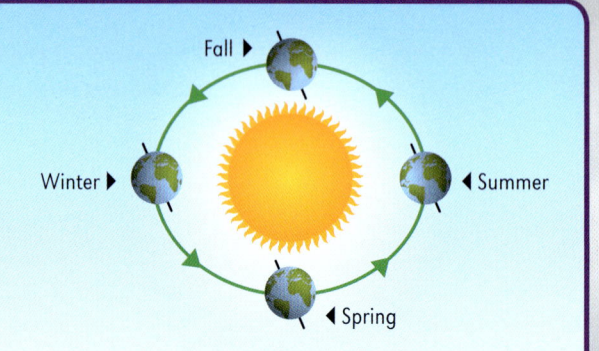

Fall ▸
◂ Winter
Summer ◂
◂ Spring

Canada is cold in the winter. This is because Canada gets indirect sunlight in the winter.

Canada is warm in the summer. This is because Canada gets direct sunlight in the summer.

Direct light has more light energy and heat energy than indirect light.

Winter — New Brunswick

Summer — New Brunswick

indirect light
direct light
sun
indirect light

indirect light
direct light
indirect light

FYI

The southern hemisphere has summer when the northern hemisphere has winter.

- northern hemisphere
- equator
- southern hemisphere

CHAPTER 2 — WEATHER PATTERNS

16

CLIMATE

Climate is what the weather is usually like in one place. The climate can be different in other places.

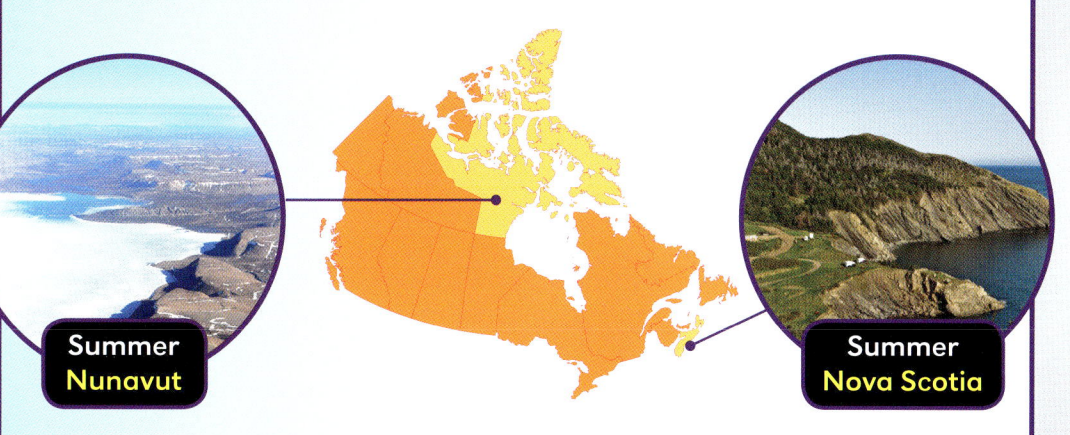

Summer Nunavut

Summer Nova Scotia

BRITISH COLUMBIA MOUNTAIN CLIMATE

Mountain climate is one type of climate.

Air with water vapour from the ocean rises up and goes over the Rocky Mountains. → The water vapour cools down as it rises up in the cool atmosphere. This causes rain or snow. → The air is dry when it gets to the other side of the mountains.

rain

water vapour

ROCKY MOUNTAINS

PACIFIC OCEAN | BRITISH COLUMBIA | ALBERTA

FYI
Places near the ocean are warmer in the winter and cooler in the summer than places far from the ocean. This is because water warms up and cools down slowly.

?
1. Why is it hot in Canada in the summer?
2. What is climate?
3. What happens to water vapour when it goes over the mountains?
4. Think about where you live. What is the climate like?

WEATHER PATTERNS — CHAPTER 2

Measuring Weather

BIG IDEA — Weather can be measured.

Scientists have invented instruments to measure weather. A thermometer is one weather instrument. It can be used to measure temperature.

You can measure the weather too. You can use your observations and measurements to predict the weather.

rain gauge

Precipitation

A rain gauge measures the amount of rain that falls.

thermometer

Temperature

A thermometer measures how hot or cold the air is.

Air Pressure

A barometer measures air pressure. You can read the air pressure on the barometer. Later, you can read the air pressure again. If the air pressure has changed, the weather is likely going to change.

barometer

Wind Speed

An anemometer measures how strong the wind is. This is called the wind speed. The wind makes the red cups spin.

anemometer

Wind Direction

A weather vane shows where the wind is coming from. Where the wind is coming from is called the wind direction. The arrow points in the direction the wind is coming from.

weather vane

FYI

A windsock shows wind direction and wind speed. The large end of the sock points in the direction the wind is coming from. A windsock pointing straight out tells you that the wind is strong.

windsock

1. Which weather instrument measures wind speed?
2. Which weather instrument do you use most? Why?
3. Which weather instrument would you like to use?

CHAPTER 2 — WEATHER PATTERNS

ACTIVITY 3: MAKE A WINDSOCK

Make a windsock to help you observe wind direction and wind speed.

YOU WILL NEED: Styrofoam cup, plastic bag, string, tape, elastic band, scissors, pencil and paper

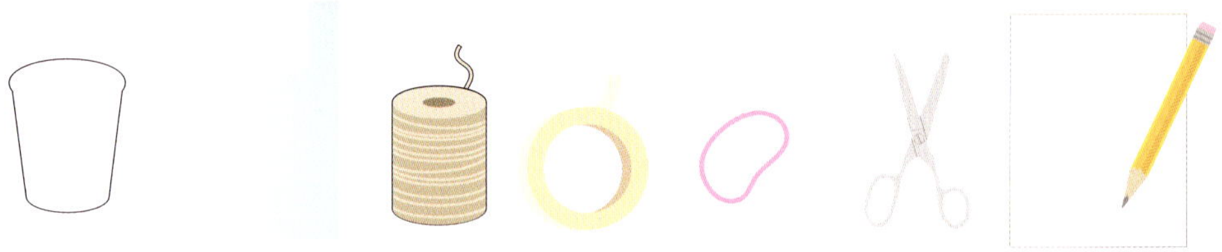

1. Cut off the bottom of the cup.

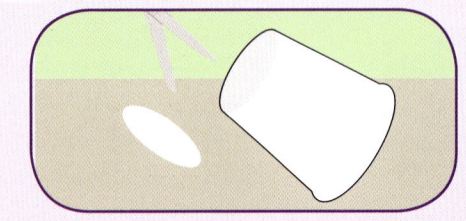

2. Use the pencil to make four holes near the top of the cup.

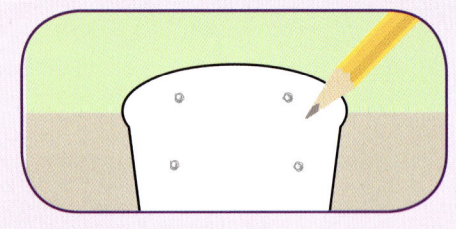

3. Cut off the bottom of the plastic bag.

4. Attach the plastic bag to the cup. Use tape or the elastic band.

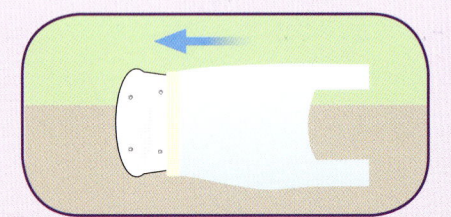

5. Tie some pieces of string to the holes in the cup.

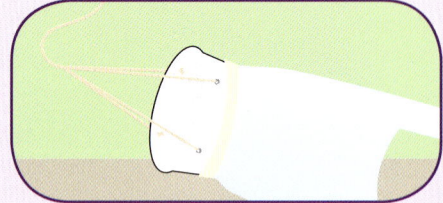

6. Put your windsock outside. Attach it to a flagpole or fence. The wind will blow into the big end of the cup.

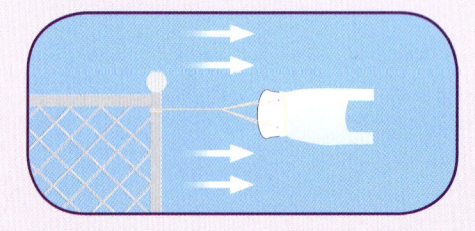

7. What direction is the wind coming from? Is it coming from the north, south, east, or west?
How strong is the wind?
Record your observations.

ANALYZE & REFLECT

1. Is the windsock a good way to measure wind direction and wind speed? Why?
2. Compare your observations of your windsock with the weather forecast for the day. How are they the same? How are they different?

20

ACTIVITY 4: KEEP A WEATHER LOG

Observe and measure the weather every day for five days.

1 Make a weather log like the one below.

2 Observe and measure the weather at the same time every day.

3 Every day, use your observations and measurements to predict the weather for later in the day. Talk about your predictions with a classmate.

	Sample	Monday	Tuesday	Wednesday	Thursday	Friday
Temperature	-2°C					
Rain	2 mm					
Snow	10 mm					
Wind Direction and Wind Speed	NE wind Light wind					
Cloud Cover	Partly Cloudy					

FYI
Sometimes the wind comes from the northeast (NE), northwest (NW), southeast (SE), or southwest (SW).

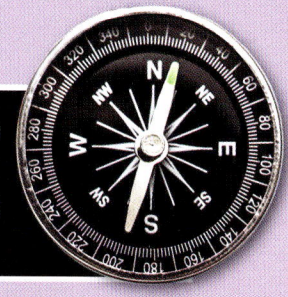

ANALYZE & REFLECT

1. Which weather observation or measurement helped you to predict rain or snow? Why?
2. How could you measure snowfall?
3. Compare your weather log to the weather report in the newspaper, on TV, or online each day. Is it the same? Why might it be different?

CHAPTER 3

WEATHER AND OUR

People have to wear the right clothes for the weather.

People do different activities in different types of weather.

snowshoeing

bodyboarding

WORLD

BIG IDEA The weather changes our world and how we live.

People build homes for the climate where they live.

FYI

Humans are changing the climate of Earth. We make gases that go into the atmosphere. Some of these gases make the atmosphere trap more heat. This is causing climate change.

sun

escaping heat

trapped heat

GASES IN THE ATMOSPHERE

Earth

1. What clothes do you need to help you live in the weather in your area?

2. What activities do you do in the winter?

3. How does your home stay warm in the winter?

Weather and Living and Non-living Things

BIG IDEA — Weather can change living and non-living things.

Plants and animals have to live with the weather. They all have to survive in different kinds of weather.

LIVING THINGS

ANIMALS

chipmunk

Some animals sleep through the winter. This is called hibernation.

Canada geese

Some birds fly south to warmer weather in the fall. This is called migration.

Some animals grow extra fur in the winter. This extra fur keeps them warm.

bison

Arctic fox

Some animals grow white fur in the winter. The white fur helps them hide in the snow. This helps them hunt and hide from other animals.

PLANTS

milkweed seeds

Some plants die in the winter. But their seeds can survive over the winter. The seeds grow in the spring.

tree

dry grass

Some plants look dead in the winter. They also look dead when there is not enough water. But their roots are still alive in the ground. The plants grow again when the weather changes.

NON-LIVING THINGS

The weather can change non-living things. Rocks and soil are non-living things. So are houses and buildings.

ice

Sometimes water freezes in a crack in a rock. This can break the rock apart.

erosion

Wind and water can slowly break rocks apart and wash away soil. This is called erosion.

hail damage

Hail can damage cars and houses. It can also damage plants.

1. What are two ways plants survive the winter?
2. What are two ways animals survive the winter?
3. How can water change non-living things?

25

Extreme Weather

BIG IDEA Some weather can harm living things and non-living things.

Some weather can cause big problems. This type of weather is called extreme weather.

Weather forecasts can help people get ready for extreme weather.

Thunderstorm

A thunderstorm is a storm with rain, thunder, and lightning. Lightning can start forest fires and destroy buildings. Lightning can also hit people and animals. Stay inside when there is lightning!

lightning

tornado

Tornado

A tornado is a storm with strong winds that spin around in a circle. Tornadoes can destroy homes and buildings.

FYI
Environment Canada measures the weather and makes weather predictions. It sends out warnings for extreme weather.

Blizzard

A blizzard is a storm with snow and strong winds. The wind makes it feel very cold. Blowing snow makes it hard to see. It is dangerous to drive during a blizzard.

blizzard

Flood

A lot of rain can cause a flood. Sometimes in the spring, melting ice in rivers can cause flooding. Floods can damage homes and buildings.

flood

Drought

A drought is a long time with no rain. There is not enough water for plants and animals during a drought. Plants and animals may die.

drought

1. How can you stay safe during a thunderstorm?
2. Which extreme weather is the most dangerous for people? Why?
3. Talk to a classmate about an extreme weather event you know about.

ACTIVITY 5: CLIMATE CHANGE

Changes in Earth's atmosphere are causing climate changes.

1 Read the chart below.

CLIMATE CHANGES

1. The temperature of Earth's atmosphere is going up.
2. There are more extreme weather events.
3. Some places get more precipitation.
4. Some places get less precipitation.

EFFECTS OF CLIMATE CHANGES

a. Ice at the North Pole and South Pole might melt.
b. Land near lakes and oceans might become flooded.
c. Some types of plants and animals might die and disappear from Earth.
d. Some people might not have enough water to drink.
e. Homes for people and animals could be destroyed.
f. There could be floods.
g. There could be droughts.
h. People might have to move.
i. Some plants might not grow well.

2 Make a list of the four climate changes.

3 Beside each climate change, list one effect that it might cause.

ANALYZE & REFLECT

1. Which climate changes have more than one effect?
2. Which effect is the most important? Talk about your answer with a classmate.
3. Talk with a classmate about each effect caused by climate change. Talk about why the effect would happen.

ACTIVITY 6: TEST CLOTH

People have clothes to protect them from the weather. Some cloth is good for stopping water. Work with a partner to find out which cloth is good for rainy days.

YOU WILL NEED
- aluminum pan 
- plastic cups 
- elastic bands 
- cotton cloth, wool cloth, nylon cloth 
- water 
- measuring cup 

1 Make a chart like the one below.

Cloth	How fast does the water go through the cloth?	How much water goes into the cup?	How wet is the cloth?
cotton			
wool			
nylon			

2 Put the piece of cotton cloth over a cup. Put an elastic band around it.

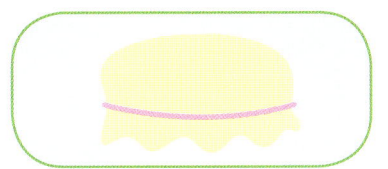

3 Put the cup on the aluminum pan.

4 Slowly pour 250 ml of water over the cloth.

5 Record your observations on your chart.

6 Repeat steps 2–5 for the wool cloth.

7 Repeat steps 2–5 for the nylon cloth.

8 Which cloth would you wear on a rainy day? Why?

ANALYZE & REFLECT

1. Was this a fair test? Explain why.
2. What protects animals from rain?
3. How could you test cloth to see if it would be good on a windy day?

CONNECTIONS

THINK LIKE A METEOROLOGIST
TO COMPARE WEATHER

Weather is different in different places in Canada. It is also different around the world. Work with a partner to compare weather in three different places.

1 Compare the weather where you live to the weather in
   a) another place in Canada
   b) another place in the world

2 Find the weather report for today for each place. You can use a newspaper or go online.

3 Make a chart like the one below. Record the information you find on your chart. You can use words, pictures, or both.

4 Talk about the weather in each place. Is it different or the same? Why?

	Where I Live	Place a	Place b
High Temperature			
Low Temperature			
Precipitation (What kind and how much?)			
Wind Direction and Wind Speed			
Cloud Cover			
Air Pressure (Is it going up or going down?)			

COMMUNICATE

1 Talk about these questions with your partner.

a) What type of clothes would people wear in each place? Why?

b) What activities can people do in the weather in each place? Why?

c) Give two reasons to explain the weather in each place.

2 Make a weather report for one of the places you researched. Present your weather report to the class.

VOCABULARY

1 The word "weather" is a noun. The words "sunny," "rainy," "windy," and "snowy" can be used to describe weather. They are called adjectives. Draw pictures of things you like to do in sunny weather, rainy weather, windy weather, and snowy weather. Label your pictures.

sunny weather

2 Make a picture dictionary with five weather words from this book. Draw a picture and write a definition for each word.

Weather Word	Definition	Drawing
precipitation	Something that falls from the sky. It can be rain, snow, sleet, or hail.	

REVIEW

Talk about these questions with a classmate.

1. What two things cause changes in the weather?

2. What causes wind?

3. How does the water cycle change the weather?

4. List three types of precipitation.

5. Which type of cloud tells you that a storm might be coming?

6. Why is it hot near the equator?

7. Which weather instrument shows more than one thing about weather?

8. What do meteorologists use to predict weather?

9. List three living things. Tell how each one survives Canada's winter.

10. List three types of extreme weather.

Glossary

energy
(en-er-jee)
a force that can be used to do work and move things; light and heat are forms of energy

extreme
(ik-streem)
very large or severe

hunt
(huhnt)
try to catch another animal for food

measurements
(mezh-er-muhnts)
the size, length, or amount of something

observe
(ob-zurv)
notice or see

predict
(prih-dikt)
use your knowledge to decide what will likely happen

record
(rih-kord)
write something down

survive
(ser-vive)
keep living; stay alive

Index

A
air, 5–10, 13, 17–18
air pressure, 8–9, 18, 30
anemometer, 19
atmosphere, 6, 11, 17, 23, 28

B
barometer, 18
blizzard, 27

C
cirrus clouds, 11
climate, 16–17, 23, 28
cloud, 4, 9–11, 13–14, 21, 30–31
condensation, 10, 12
cumulonimbus clouds, 11
cumulus clouds, 11

D
drought, 27–28

E
erosion, 25
evaporation, 10, 12

F
fall, 14–16, 24
flood, 27–28

H
hail, 4, 10, 25, 31

M
meteorologist, 13–15, 30–31
mountain, 17

P
patterns, 14–15
precipitation, 10–13, 18, 28, 30–31

R
rain, 5, 9–10, 12–13, 17–18, 21, 26–27, 29, 31
rain gauge, 18

S
season, 14–16
sleet, 5, 10, 31
snow, 5, 9–10, 13, 17, 21–22, 24, 27, 31
spring, 14, 16, 25, 27
stratus clouds, 11
summer, 9, 14–17

T
temperature, 13, 18, 21, 28, 30
thermometer, 18
thunderstorm, 11, 26–27
tornado, 4, 26

W
water, 9–10, 12, 17, 25, 27–29
water cycle, 10–12, 31
weather vane, 19
wind, 4–5, 9, 19–21, 25–27, 29–31
windsock, 19–20
winter, 15–17, 23–25, 31